Machines

Push and Pull

by Linda Ward Beech

Contents

force
A **force** is a push or a pull.

This machine uses **force** to move dirt.

push

When you **push** something, you move it away from you.

This plow **pushes** the snow.

pull

When you **pull** something, you move it toward you.

This machine **pulls** logs.

direction

A **direction** is the path an object takes.

This truck turns to change **direction.**

motion

When an object is moving, it is in **motion.**

direction

force

motion

pull

push

This truck is in **motion** on the highway.

Machines Move

There are many kinds of machines. These machines are moving. They are moving sand to build a canal.

Pushes and Pulls

Machines use **force** to move things.

force

A **force** is a push or a pull.

A force is a **push** or a **pull.** This machine pushes dirt.

push

push

When you **push** something, you move it away from you.

pull

When you **pull** something, you move it toward you.

This bulldozer pushes. It piles sand high on a beach.

push

This snowplow pushes snow off the road.

Now cars can use the road.

push

A backhoe pulls up rocks and dirt.
It can dig a hole.

This machine pulls trees from one place to another. It clears land for a new road.

Pushes and pulls can make things start and stop moving. The man pushes the wheelbarrow to make it start moving.

This man moves things in a wheelbarrow.

Trucks can use force to change **direction.**
The truck cab pulls the trailer in a new direction.

direction

A **direction** is the path an object takes.

Ways Objects Move

Pushes and pulls can move machines. They can put machines in **motion.**

This truck moves fast on a smooth road.

motion

When an object is moving, it is in **motion.**

Machines push against the ground to move forward. Sometimes machines move fast. Sometimes machines move slowly.

This bulldozer moves slowly on the rough ground.

Objects can move in different ways.

A cement mixer turns round-and-round.

This man pushes and pulls the saw.
It moves back-and-forth to cut wood.

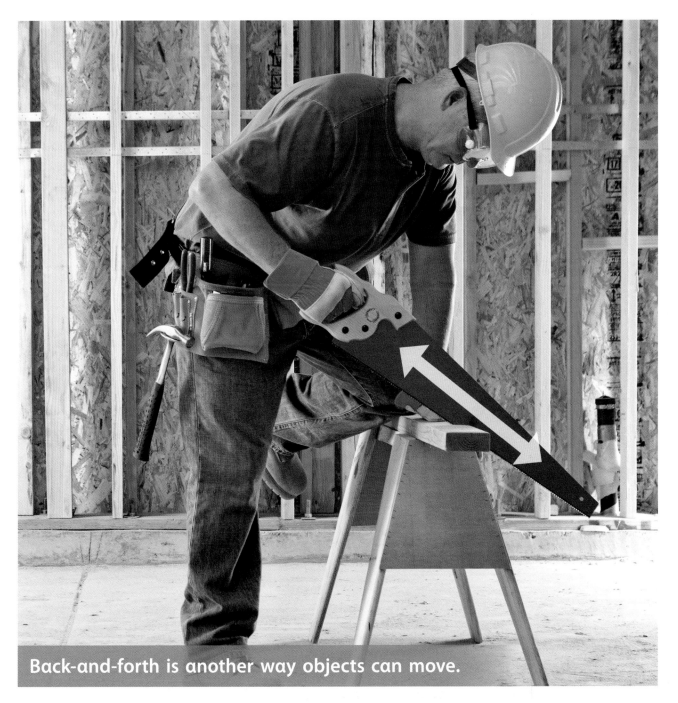

Back-and-forth is another way objects can move.

This road roller smoothes the street. It is moving in a straight line.

These cars zigzag down a city street.

zigzag

Find the Pushes and Pulls

This picture shows people using forces to move things. Name the pushes and pulls you see. Two are labeled to get you started.

pull

push

Magnets Move Objects

A magnet can make a pulling force. It can pull some metals.

magnet —

This large magnet picks up scrap metal.
It is helping to clean up a building site.

magnet ———

scrap metal

This magnet pulls objects made of iron.

Conclusion

Pushes and pulls are forces. Forces can make machines move in different directions. A magnet can pull some objects like scrap metal.

Think About the Big Ideas

1. What is a push or pull?
2. What are some ways machines move?

Share and Compare

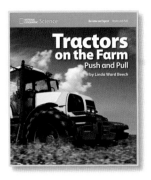 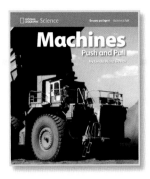 

Turn and Talk

Compare the pushes and pulls in your books. How are they alike? How are they different?

Read

Find a photo with a caption. Read the caption to a classmate.

Write

Write about how something in your book moves. Share what you wrote with a classmate.

Draw

Draw something pushing or pulling. Share your drawing with a classmate.

Meet Marianne Dyson

Marianne Dyson was one of the first women to work for NASA. She used to help crews fly the space shuttle.

Today she writes children's books about pushes and pulls on Earth and in space. She wrote a book called *Home on the Moon*. She explains forces on the moon and what it would be like to live there.

Index

Acknowledgments
Grateful acknowledgment is given to the authors, artists, photographers, museums, publishers, and agents for permission to reprint copyrighted material. Every effort has been made to secure the appropriate permission. If any omissions have been made or if corrections are required, please contact the Publisher.

Photographic Credits
Cover (bg) Paul Mayall/Photographers Direct; Cvr Flap (t), 12 Kochneva Tetyana/Shutterstock; Cvr Flap (b), 21 Juice Images/PunchStock; Title (bg) Patrick Laverdant/iStockphoto; 2-3 Joerg Reimann/iStockphoto; 4, 10-11 Pacific Stock/SuperStock; 5 (t), 13 Lloyd Paulson/Shutterstock; 5 (b), 15 Larry Lefever/Grant Heilman Photography; 6, 17 Nick Suydam/Alamy Images; 7, 18 Konstantin Sutyagin/Shutterstock; 8-9 Christine Osborne/Corbis; 14 Lukas Pobuda/Shutterstock; 16 David Young-Wolff/PhotoEdit; 19 Digital Vision/Getty Images; 20 Florin C/Shutterstock; 22 Filip Fuxa/Shutterstock; 23 Kevin Connors/Shutterstock; 26 Leonard Lessin/Peter Arnold Inc./Alamy Images; 27 Paul Ridsdale/Alamy Images; 28 Laurentiu Nica/Shutterstock; 30 The Stocktrek Corp/Brand X/Corbis; 31 Bruce Bennett/National Geographic Image Collection; Inside Back Cover (bg) lebanmax/Shutterstock.

Illustrator Credits
24-25 Adrian Chesterman.

Neither the Publisher nor the authors shall be liable for any damage that may be caused or sustained or result from conducting any of the activities in this publication without specifically following instructions, undertaking the activities without proper supervision, or failing to comply with the cautions contained herein.

Program Authors
Malcolm B. Butler, Ph.D., Associate Professor of Science Education, University of South Florida, St. Petersburg, Florida; Judith Sweeney Lederman, Ph.D., Director of Teacher Education and Associate Professor of Science Education, Department of Mathematics and Science Education, Illinois Institute of Technology, Chicago, Illinois; Randy Bell, Ph.D., Associate Professor of Science Education, University of Virginia, Charlottesville, Virginia; Kathy Cabe Trundle, Ph.D., Associate Professor of Early Childhood Science Education, The Ohio State University, Columbus, Ohio; Nell K. Duke, Ed.D., Co-Director of the Literacy Achievement Research Center and Professor of Teacher Education and Educational Psychology, Michigan State University, East Lansing, Michigan; David W. Moore, Ph.D., Professor of Education, College of Teacher Education and Leadership, Arizona State University, Tempe, Arizona

The National Geographic Society
John M. Fahey, Jr., President & Chief Executive Officer
Gilbert M. Grosvenor, Chairman of the Board

National Geographic School Publishing
Hampton-Brown
www.NGSP.com

Printed in the USA.
RR Donnelley, Atlanta, GA

ISBN: 978-0-7362-7594-1

11 12 13 14 15 16 17

10 9 8 7 6 5 4 3 2